Author

Name

Department

Email   Phone (   )

Signature

Book ID

Book Number   Start Date   /   /

Continue from book number

Final page used   End Date   /   /

Continue to book number

Project Title   Page No.

Project Title   Page No.

Project Title   Page No.

Location

Address

Department

Room

Issuer

Name

Department

Email   Phone (   )

Witness

Name

Department

Email   Phone (   )

TABLE OF CONTENTS

Book No._____   Book Start Date_____

Project No.	Project Title	Page

TABLE OF CONTENTS

Book No._____   Book Start Date _____

Project No.	Project Title	Page

Continue from page_____

Garry
Gar-bear   Woz
Wozniak
Wawa
Wozzy G
G-woz
li'l woz
little woz

Continue to page _____

Proprietary Information	Invented by:	Date	Witnessed and Understood by me	Date
	Recorded by:			

$x^x = 1$ $\log_x 1 = x$ $x = 0$

Project No._____ Project Title:_____ Page No. ___2___

Continue from page_____

Continue to page _____

Proprietary Information	Invented by:	Date	Witnessed and Understood by me	Date
	Recorded by:			

$\log(1) = 0$ $\ln(1) = 0$

Project No.＿＿＿＿＿＿ Project Title:＿＿＿＿＿＿＿＿＿＿＿＿＿＿＿

Continue from page＿＿＿＿＿

Continue to page＿＿＿＿＿

Proprietary Information	Invented by:	Date	Witnessed and Understood by me	Date
	Recorded by:			

Project No._____ Project Title:_____ Page No. _____4_____

Continue from page_____

Continue to page _____

Proprietary Information	Invented by:		Date	Witnessed and Understood by me	Date
	Recorded by:				

Project No._____ Project Title:_____ Page No. ___5___

Continue from page_____

Continue to page _____

Proprietary Information	Invented by:		Date	Witnessed and Understood by me	Date
	Recorded by:				

Project No._____ Project Title:_____ Page No. ____6____

Continue from page_____

Continue to page _____

Proprietary Information	Invented by:	Date	Witnessed and Understood by me	Date
	Recorded by:			

Continue from page_____

Continue to page _____

| Proprietary Information | Invented by: | Date | Witnessed and Understood by me | Date |
| | Recorded by: | | | |

Continue from page_____

Continue to page _____

Proprietary Information	Invented by:	Date	Witnessed and Understood by me	Date
	Recorded by:			

Continue from page_____

Continue to page _____

Proprietary Information	Invented by:	Date	Witnessed and Understood by me	Date
	Recorded by:			

Project No._____ Project Title:_____ Page No. __10__

Continue from page_____

Continue to page _____

Proprietary Information	Invented by:		Date	Witnessed and Understood by me	Date
	Recorded by:				

Project No._____ Project Title:_____

Continue from page_____

Continue to page _____

Proprietary Information	Invented by:		Date	Witnessed and Understood by me	Date
	Recorded by:				

Project No._____ Project Title:_____ Page No. ___12___

Continue from page_____

Continue to page _____

| Proprietary Information | Invented by: | Date | Witnessed and Understood by me | Date |
| | Recorded by: | | | |

Continue from page_____

Continue to page _____

Proprietary Information	Invented by:		Date	Witnessed and Understood by me	Date
	Recorded by:				

Continue from page_____

Continue to page _____

Proprietary Information	Invented by:		Date	Witnessed and Understood by me	Date
	Recorded by:				

Continue from page_____

Continue to page _____

Proprietary Information	Invented by:		Date	Witnessed and Understood by me	Date
	Recorded by:				

Project No._____ Project Title:_____ Page No. __16__

Continue from page_____

Continue to page _____

| Proprietary Information | Invented by: | Date | Witnessed and Understood by me | Date |
| | Recorded by: | | | |

Continue from page_____

Continue to page _____

Proprietary Information	Invented by:	Date	Witnessed and Understood by me	Date
	Recorded by:			

Project No._____ Project Title:_____ Page No. <u>18</u>

Continue from page_____

Continue to page _____

Proprietary Information	Invented by:		Date	Witnessed and Understood by me	Date
	Recorded by:				

Continue from page_____

Continue to page _____

Proprietary Information	Invented by:	Date	Witnessed and Understood by me	Date
	Recorded by:			

Continue from page_____

Continue to page _____

Proprietary Information	Invented by:	Date	Witnessed and Understood by me	Date
	Recorded by:			

Project No._____ Project Title:_____ Page No. __21__

Continue from page_____

Continue to page _____

| Proprietary Information | Invented by: | Date | Witnessed and Understood by me | Date |
| | Recorded by: | | | |

Project No._____ Project Title:_____ Page No. __22__

Continue from page_____

Continue to page _____

| Proprietary Information | Invented by: | | Date | Witnessed and Understood by me | Date |
| | Recorded by: | | | | |

Project No._____ Project Title:_____ Page No. __23__

Continue from page_____

Continue to page _____

Proprietary Information	Invented by:	Date	Witnessed and Understood by me	Date
	Recorded by:			

Project No._____ Project Title:_____ Page No. <u>24</u>

Continue from page_____

Continue to page _____

Proprietary Information	Invented by:		Date	Witnessed and Understood by me	Date
	Recorded by:				

Continue from page_____

Continue to page _____

Proprietary Information	Invented by:	Date	Witnessed and Understood by me	Date
	Recorded by:			

Project No._____ Project Title:_____ Page No. _26___

Continue from page_____

Continue to page _____

Proprietary Information	Invented by:		Date	Witnessed and Understood by me	Date
	Recorded by:				

Project No._____ Project Title:_____

Continue from page_____

Continue to page _____

| Proprietary Information | Invented by: | Date | Witnessed and Understood by me | Date |
| | Recorded by: | | | |

Project No._____ Project Title:_____ Page No. __28__

Continue from page_____

Proprietary Information	Invented by:	Date	Witnessed and Understood by me	Date
	Recorded by:			

Continue from page＿＿＿＿＿

Continue to page ＿＿＿＿＿

Proprietary Information	Invented by:	Date	Witnessed and Understood by me	Date
	Recorded by:			

Continue from page_____

Continue to page _____

Proprietary Information	Invented by:	Date	Witnessed and Understood by me	Date
	Recorded by:			

Project No._____ Project Title:_____

Continue from page_____

Continue to page _____

Proprietary Information	Invented by:	Date	Witnessed and Understood by me	Date
	Recorded by:			

Continue from page_____

Continue to page _____

Proprietary Information	Invented by:	Date	Witnessed and Understood by me	Date
	Recorded by:			

Continue from page_____

Continue to page _____

Proprietary Information	Invented by:		Date	Witnessed and Understood by me	Date
	Recorded by:				

Continue from page_____

Continue to page _____

Proprietary Information	Invented by:	Date	Witnessed and Understood by me	Date
	Recorded by:			

Continue from page_____

Continue to page _____

Proprietary Information	Invented by:	Date	Witnessed and Understood by me	Date
	Recorded by:			

Continue from page_____

Continue to page _____

Proprietary Information	Invented by:		Date	Witnessed and Understood by me	Date
	Recorded by:				

Project No._____ Project Title:_____ Page No. _37_

Continue from page_____

Continue to page _____

Proprietary Information	Invented by:	Date	Witnessed and Understood by me	Date
	Recorded by:			

Continue from page_____

Continue to page _____

Proprietary Information	Invented by:	Date	Witnessed and Understood by me	Date
	Recorded by:			

Continue from page_____

Continue to page _____

| Proprietary Information | Invented by: | Date | Witnessed and Understood by me | Date |
| | Recorded by: | | | |

Continue from page_____

Continue to page _____

Proprietary Information	Invented by:	Date	Witnessed and Understood by me	Date
	Recorded by:			

Project No._____ Project Title:_____ Page No. _41_

Continue from page_____

Continue to page _____

| Proprietary Information | Invented by:_____ | Date | Witnessed and Understood by me | Date |
| | Recorded by: | | | |

Continue from page_____

Continue to page _____

Proprietary Information	Invented by:	Date	Witnessed and Understood by me	Date
	Recorded by:			

Project No._____ Project Title:_____ Page No. __43__

Continue from page_____

Continue to page _____

Proprietary Information	Invented by:	Date	Witnessed and Understood by me	Date
	Recorded by:			

Continue from page_____

Continue to page _____

Proprietary Information	Invented by:		Date	Witnessed and Understood by me	Date
	Recorded by:				

Project No._____ Project Title:_____ Page No. 45

Continue from page_____

Continue to page _____

Proprietary Information	Invented by:	Date	Witnessed and Understood by me	Date
	Recorded by:			

Continue from page_____

Continue to page _____

Proprietary Information	Invented by:		Date	Witnessed and Understood by me	Date
	Recorded by:				

Continue from page_____

Continue to page _____

Proprietary Information	Invented by:	Date	Witnessed and Understood by me	Date
	Recorded by:			

Project No._____ Project Title:_____ Page No. __48__

Continue from page_____

Continue to page _____

| Proprietary Information | Invented by: | | Date | Witnessed and Understood by me | Date |
| | Recorded by: | | | | |

Continue from page_____

Continue to page _____

Proprietary Information	Invented by:	Date	Witnessed and Understood by me	Date
	Recorded by:			

Project No._____ Project Title:_____ Page No. 50

Continue from page_____

Continue to page _____

Proprietary Information	Invented by:		Date	Witnessed and Understood by me	Date
	Recorded by:				

Continue from page_____

Continue to page _____

Proprietary Information	Invented by:	Date	Witnessed and Understood by me	Date
	Recorded by:			

Continue from page_____

Continue to page _____

Proprietary Information	Invented by:	Date	Witnessed and Understood by me	Date
	Recorded by:			

Project No._____ Project Title:_____

Continue from page_____

Continue to page _____

Proprietary Information	Invented by:	Date	Witnessed and Understood by me	Date
	Recorded by:			

Continue from page_____

Continue to page _____

Proprietary Information	Invented by:	Date	Witnessed and Understood by me	Date
	Recorded by:			

Project No._____ Project Title:_____

Continue from page_____

Continue to page _____

Proprietary Information	Invented by:	Date	Witnessed and Understood by me	Date
	Recorded by:			

Continue from page_____

Continue to page _____

Proprietary Information	Invented by:		Date	Witnessed and Understood by me	Date
	Recorded by:				

Continue from page_____

Continue to page _____

Proprietary Information	Invented by:	Date	Witnessed and Understood by me	Date
	Recorded by:			

Continue from page_____

Continue to page _____

Proprietary Information	Invented by:	Date	Witnessed and Understood by me	Date
	Recorded by:			

Continue from page_____

Continue to page _____

Proprietary Information	Invented by:	Date	Witnessed and Understood by me	Date
	Recorded by:			

Continue from page_____

Continue to page _____

Proprietary Information	Invented by:		Date	Witnessed and Understood by me	Date
	Recorded by:				

Project No._____ Project Title:_____

Continue from page_____

Continue to page _____

Proprietary Information	Invented by:	Date	Witnessed and Understood by me	Date
	Recorded by:			

Project No._____ Project Title:_____ Page No. _62_

Continue from page_____

Continue to page _____

Proprietary Information	Invented by:	Date	Witnessed and Understood by me	Date
	Recorded by:			

Continue from page _____

Continue to page _____

Proprietary Information	Invented by:	Date	Witnessed and Understood by me	Date
	Recorded by:			

Continue from page_____

Continue to page _____

Proprietary Information	Invented by:	Date	Witnessed and Understood by me	Date
	Recorded by:			

Project No._____ Project Title:_____ Page No. 65

Continue from page_____

Continue to page _____

| Proprietary Information | Invented by: | Date | Witnessed and Understood by me | Date |
| | Recorded by: | | | |

Continue from page_____

Continue to page _____

Proprietary Information	Invented by:	Date	Witnessed and Understood by me	Date
	Recorded by:			

Project No._____ Project Title:_____ Page No. __67__

Continue from page_____

Continue to page _____

Proprietary Information	Invented by:	Date	Witnessed and Understood by me	Date
	Recorded by:			

Continue from page_____

Continue to page _____

Proprietary Information	Invented by:	Date	Witnessed and Understood by me	Date
	Recorded by:			

Project No._____ Project Title:_____

Continue from page_____

Continue to page _____

Proprietary Information	Invented by:	Date	Witnessed and Understood by me	Date
	Recorded by:			

Continue from page_____

Continue to page _____

Proprietary Information	Invented by:	Date	Witnessed and Understood by me	Date
	Recorded by:			

Continue from page_____

Continue to page _____

Proprietary Information	Invented by:	Date	Witnessed and Understood by me	Date
	Recorded by:			

Continue from page_____

Continue to page _____

| Proprietary Information | Invented by: | Date | Witnessed and Understood by me | Date |
| | Recorded by: | | | |

Project No._____ Project Title:_____ Page No. __73__

Continue from page_____

Continue to page _____

Proprietary Information	Invented by:	Date	Witnessed and Understood by me	Date
	Recorded by:			

Project No._____ Project Title:_____

Continue from page_____

Continue to page _____

Proprietary Information	Invented by:		Date	Witnessed and Understood by me	Date
	Recorded by:				

Project No._____ Project Title:_____ Page No. _75_

Continue from page_____

Continue to page _____

Proprietary Information	Invented by:	Date	Witnessed and Understood by me	Date
	Recorded by:			

Continue from page_____

Continue to page _____

Proprietary Information	Invented by:	Date	Witnessed and Understood by me	Date
	Recorded by:			

Project No._____ Project Title:_____

Continue from page_____

Continue to page _____

Proprietary Information	Invented by:	Date	Witnessed and Understood by me	Date
	Recorded by:			

Continue from page_____

Continue to page _____

Proprietary Information	Invented by:		Date	Witnessed and Understood by me	Date
	Recorded by:				

Continue from page_____

Continue to page _____

| Proprietary Information | Invented by: | Date | Witnessed and Understood by me | Date |
| | Recorded by: | | | |

Continue from page_____

Continue to page _____

Proprietary Information	Invented by:		Date	Witnessed and Understood by me	Date
	Recorded by:				

Continue from page_____

Continue to page _____

Proprietary Information	Invented by:	Date	Witnessed and Understood by me	Date
	Recorded by:			

Project No._____ Project Title:_____ Page No. __82__

Continue from page_____

Continue to page _____

Project No._____ Project Title:_____

Continue from page_____

Continue to page _____

Proprietary Information	Invented by:	Date	Witnessed and Understood by me	Date
	Recorded by:			

Continue from page_____

Continue to page _____

Proprietary Information	Invented by:		Date	Witnessed and Understood by me	Date
	Recorded by:				

Continue from page_____

Continue to page _____

Proprietary Information	Invented by:	Date	Witnessed and Understood by me	Date
	Recorded by:			

Continue from page_____

Continue to page _____

Proprietary Information	Invented by:		Date	Witnessed and Understood by me	Date
	Recorded by:				

Continue from page_____

Continue to page _____

Proprietary Information	Invented by:	Date	Witnessed and Understood by me	Date
	Recorded by:			

Continue from page_____

Continue to page _____

Proprietary Information	Invented by:		Date	Witnessed and Understood by me	Date
	Recorded by:				

Project No._____ Project Title:_____

Continue from page_____

Continue to page _____

Proprietary Information	Invented by:	Date	Witnessed and Understood by me	Date
	Recorded by:			

Continue from page_____

Continue to page _____

Proprietary Information	Invented by:		Date	Witnessed and Understood by me	Date
	Recorded by:				

Continue from page_____

Continue to page _____

Proprietary Information	Invented by:	Date	Witnessed and Understood by me	Date
	Recorded by:			

Continue from page_____

Continue to page _____

Proprietary Information	Invented by:	Date	Witnessed and Understood by me	Date
	Recorded by:			

Project No._____ Project Title:_____

Continue from page_____

Continue to page _____

Proprietary Information	Invented by:	Date	Witnessed and Understood by me	Date
	Recorded by:			

Continue from page_____

Continue to page _____

Proprietary Information	Invented by:		Date	Witnessed and Understood by me	Date
	Recorded by:				

Project No._____ Project Title:_____ Page No. <u>95</u>

Continue from page_____

Continue to page _____

| Proprietary Information | Invented by: | Date | Witnessed and Understood by me | Date |
| | Recorded by: | | | |

Continue from page_____

Continue to page _____

Proprietary Information	Invented by:		Date	Witnessed and Understood by me	Date
	Recorded by:				

Made in the USA
Columbia, SC
28 September 2018